Apollo 1 Tragedy

Other titles in the *American Disasters* series:

Apollo 1 Tragedy

Fire in the Capsule

Paul Brubaker

Enslow Publishers, Inc.

40 Industrial Road	PO Box 38
Box 398	Aldershot
Berkeley Heights, NJ 07922	Hants GU12 6BP
USA	UK

http://www.enslow.com

To my loving wife, Lynette

Copyright © 2002 by Enslow Publishers, Inc.

Library of Congress Cataloging-in-Publication Data

Brubaker, Paul.
    Apollo 1 tragedy : fire in the capsule / Paul Brubaker.
        p. cm. — (American disasters)
    Includes bibliographical references and index.
    Summary: An account of the Apollo 1 fire that claimed the lives of three
astronauts in 1967.
    ISBN 0-7660-1787-7
    1. Project Apollo (U.S.)—Juvenile literature. 2. Space vehicle accidents—United
States—Juvenile literature. [1. Project Apollo (U.S.) 2. Space vehicle accidents.] I.
Title. II. Series.
    TL789.8.U6 A52395 2002
    363.12'4—dc21
                          2001004018

Printed in the United States of America

10 9 8 7 6 5 4 3 2 1

Illustration Credits: National Aeronautics and Space Administration (NASA).

Cover Illustration: National Aeronautics and Space Administration (NASA).

Contents

*T*he *Apollo 1* capsule is lifted to the top of the gantry during tests prior to its scheduled launch.

CHAPTER 1

The Strange Day of January 27, 1967

"**H**ow are we going to get to the moon if we can't talk between three buildings?" snapped a frustrated Commander Grissom from the cockpit of the *Apollo 1* spacecraft.[1] A voice responded but all he could hear in the headphones in his helmet was crackling static. "I can't hear a thing you're saying," said Grissom. Finally a voice came through, asking "did you say something in the command module?" Grissom repeated, "I said, 'I don't know how we're going to get to the moon if we can't talk between two or three buildings.'"[2] More static filled his ears. The commander was no longer annoyed, he was just plain angry.

It had been a long frustrating day for everyone working at Kennedy Space Center on January 27, 1967. A thousand people had been working since 8:00 A.M. testing the different parts of the new *Apollo 1* spacecraft. The *Apollo 1* was designed to go to the moon.[3]

In January 1967, the moon did not seem very far away from the United States. The experts in the nation's space program had learned a tremendous amount about space exploration in a very short time. Project Apollo, the series of three-man missions that would lead to a man walking on the moon, was preparing to launch its first mission. Astronauts Virgil I. "Gus" Grissom, Edward White, and Roger Chaffee were chosen as the crew.

The *Apollo 1* was unlike any spacecraft NASA had ever designed before. The huge Saturn 1-B booster rocket stood 224 feet tall on launchpad 34, but it contained no fuel.[4] On top of the rocket was the new space command module which was the most complex vehicle ever created by NASA. Because it was so complex, there were many problems with the new spacecraft. None of the tests being done on January 27 were considered dangerous, but nearly all of them found some kind of glitch. It was not until about 1:00 P.M. that all of the many technicians and scientists were ready for the crew of *Apollo 1* to board the spacecraft for the final test.[5]

The problems continued when the crew got on board. The internal oxygen was switched on and Grissom smelled an odor that was like sour milk. Something was wrong with the environmental system.[6] Once the air was cleaned, a long, frustrating afternoon that would end in tragedy began for Commander Virgil I. "Gus" Grissom and astronauts Edward White and Roger Chaffee.

The last chore was to be a "plugs out" test. This was a dress rehearsal of the launch. *Apollo 1* was scheduled to lift off on February 21, 1967. The mission was to take this spacecraft on a test flight. The "plugs out" test was to give

*A*stronauts Roger Chaffee, Ed White, and Gus Grissom are seated in the *Apollo 1* capsule during a "plugs out" test.

the crew and everyone working on the team a chance to work in the same conditions as those in the actual launch. The astronauts wore their space suits. A countdown clock was used. The command module was filled with pure pressurized oxygen and the hatch sealed.[7]

Everyone and everything was supposed to work as if it was the day of the launch. Instead, it was a long day of little things going wrong. As the day went on, people grew tired. The astronauts sat in the couches for more than five hours and the test was still not completed. To make matters worse, Grissom was having trouble communicating with the other buildings of the space center.

The static that was interfering with Grissom's communication was caused by electricity.[8] The command module was a big network of systems that had to work together. The systems were all powered by electricity and linked by massive amounts of wiring. To make the conditions of the "plugs out" test even more like the actual launch, the command module had to switch from running on the huge batteries it was connected to on the launch pad to its own electrical power. That switch was to take place when the countdown reached zero.[9]

At about ten minutes before the countdown ended, the flight director stopped the clock. There was a last effort to correct the communications problems.[10] While everyone was working and the astronauts were waiting, a sparked jumped out somewhere in the thirty miles of electrical wire. No one knew about the spark until it was too late.

At 6:31 P.M. cries were heard from the command module. "We've got a fire in the cockpit! We've got a bad fire. . .get us out. We're burning up!"[11] The spark, in an atmosphere of pure oxygen, set off an inferno that engulfed the three astronauts in flames.[12] They struggled to open the hatch, but the pressure of the oxygen forced the hatch closed.[13] Outside the command module the technicians from the white room battled the intense heat, poisonous smoke, and two explosions as they tried to free the trapped astronauts.[14] Emergency crews raced to the launchpad. Some thought that the astronauts in their space suits might survive the fire.[15]

When the hatch was finally opened, almost six minutes after the blaze had started, the awful truth was that all three astronauts had died.[16] "Most of their suits were still white. You didn't see three charred bodies inside," said Stu Roosa who was working in the operations building and raced to the launchpad.[17] A doctor later determined that the three men died from breathing the poisonous gases that came from the burning materials in the spacecraft.[18]

*G*us Grissom, Ed White and Roger Chaffee lost their lives in the *Apollo 1* fire on January 27, 1967.

The *Apollo 1* fire had claimed three lives. These astronauts had not died in flight or in space but on the ground, a fact which was hard for many people in the space program to accept. It was strange and tragic that Grissom, White, and Chaffee should lose their lives while sitting still during a long, boring test.

With their loss, the moon, which had seemed so close, suddenly felt further away than ever before.

CHAPTER 2

The Space Race

Apollo 1 had been scheduled to be launched in February 1967, on a test flight in space.[1] The fact that NASA believed it was ready to accomplish such a feat in 1967 was a victory in itself. Not long before, the United States seemed to be losing the space race.

The space race was born out of the Cold War between the United States and the Soviet Union. As enemies during the Cold War, both nations feared each other's beliefs, influence, and military strength. That is why the United States was anxious when the Soviet Union launched the first man-made satellite called *Sputnik* (which means "satellite" in Russian) on October 4, 1957.[2] It was only the size of a large beach ball, and it emitted a small beeping sound. Less than a month later, the Russians launched *Sputnik II* into orbit. This satellite carried a passenger, a small dog named Laika. As American citizens thought about Soviet satellites flying overhead, many feared the

Russians would someday drop a bomb from one of their satellites.[3]

President Eisenhower answered the Soviet Union's orbiters by starting a new government agency in 1958, the National Aeronautics and Space Administration, commonly known as NASA. Eisenhower gave NASA the job of sending a man into space and returning him home.[4] That meant that NASA had to develop rockets and recruit astronauts.

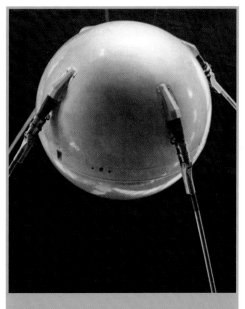

*T*he first man-made satellite, *Sputnik*, launched by the Soviet Union on October 4, 1957.

The first American attempts at building rockets were disastrous. When the United States tried to launch its own *Vanguard* satellite, the rocket barely got off the ground. After a few seconds, it toppled over and burst into flames.[5]

NASA was more successful in finding astronauts. One hundred and ten test pilots applied as volunteers for the new space program. After many demanding tests, NASA selected seven men to be the nation's first astronauts. They became known as the Mercury Seven and they would work on NASA's first mission, Project Mercury. The goal of the project was M.I.S.S., which stood for "man in space soonest."[6]

Unfortunately for NASA, the Russians put their man in space sooner. He was Yuri Gargarin, a strong, brash young man who orbited the earth in eighty-nine minutes on April 12, 1961.[7]

About two weeks later, on May 5, the United States put a man in space. Alan Shepard was just as strong, and had a cocky sense of humor. His flight, however, was shorter, lasting only fifteen minutes, and he did not orbit the earth.[8]

*A*lan B. Shepard was the first American in space. His flight launched on May 5, 1961, and it lasted fifteen minutes.

Nevertheless, Shepard's flight meant a lot to the United States and its space program. It proved that the nation could compete in the space race. The Russians may have been ahead, but the Americans were not far behind. The flight excited the American public, including the new president, John F. Kennedy.

Not long after Shepard's flight, Kennedy issued a challenge to the United States that was greater than anything the Soviet Union had proposed. "I believe that this nation should commit itself to achieving the goal, before this decade is out, of landing a man on the moon and returning him safely to the earth," said Kennedy.[9] In order for NASA to meet Kennedy's challenge, they would have to be better at space exploration than the Russians. Furthermore, Kennedy's challenge put a new competitor in the space race. The space race was no longer only against the Russians. It became a race against time.

NASA fully committed itself to meeting Kennedy's challenge. Project Mercury, the series of one-man missions and the first stage in getting to the moon, produced many successes. One of the most celebrated missions of Project Mercury was that of John Glenn orbiting the earth three times in the space capsule *Friendship 7.*

*I*n 1961, President John F. Kennedy challenged the United States to commit itself to putting a man on the moon before the end of the decade.

Once NASA could put a man in space, it began a series of two-man missions called Project Gemini. More astronauts were recruited into the space program and different missions were undertaken. Astronauts were put into space for longer periods of time to see what the effects would be. Astronauts proved they could survive and work outside of their spacecrafts during the first extra-vehicular activities (EVA), or space walks. Also, astronauts learned how to link spacecrafts together in a maneuver called a rendezvous, which would be the key to bringing men home from the moon.

By the time Project Gemini had been completed at the end of 1966, the space program had been enormously successful. The astronauts and scientists working in NASA had learned an incredible amount about space exploration and space travel in a very short time. They were even starting to believe that they might be ahead of the Russians in the space race. As 1967 began, everyone was aware that there were only three more years left to meet President Kennedy's deadline.

NASA employees were getting excited. They had proved that they could do everything they needed to do to get to the moon, walk on it, and come home. It was time to start Project Apollo, the series of missions that would lead to the first moon walk. All they had to do was test the spacecraft and get going. "Go fever" had set in at NASA. Everyone was so feverishly excited to go to the moon that the risk of an accident was very real.[10]

CHAPTER 3

The Three Men of Apollo 1

"Most people who know me know I'm not the hero type," said Virgil I. "Gus" Grissom to *Life* magazine in 1961. "There's a big difference between [astronauts] and Columbus and Lindbergh and the Wright brothers and all these people we're compared to. They did it themselves. We didn't think up this thing. We're just going to ride the capsule."[1] Indeed, Grissom did not become an astronaut to be a hero. He did believe it was important work for the nation. He also felt that it was the logical next step in his career.

Grissom's roots, like his attitude, might be considered humble by some. His childhood was spent in Mitchell, Indiana. Grissom remembered it being "as American as blueberry pie." He was the oldest of four children. His father worked for the railroad. His grades in school were average and his interest in flying was limited to building a couple of model airplanes as a child.

That changed when Grissom was a teenager and World War II raged in Europe. "The war in Europe seemed

*T*he three astronauts of *Apollo 1* (from left to right): Virgil I. "Gus" Grissom, Edward H. White II, and Roger B. Chaffee.

very far away from Indiana," remembered Grissom. "Certainly flying sounded a lot more exciting than walking. Then came Pearl Harbor, and I decided it would be the Air Force for me as soon as I finished high school and could try to qualify as an aviation cadet."[2] The war ended, however, before Grissom got his flight training.

It was during the next war, the conflict in Korea, when Grissom became a fighter pilot. By the time the Korean War began, Grissom had earned a mechanical engineering degree at Purdue University and had re-enlisted in the United States Air Force. During the war, Grissom flew one hundred combat missions with the 334th Fighter Interceptor Squadron.[3] When the war ended, Grissom remained with the air force as a test pilot and achieved the rank of captain.

During this time Grissom got a strange message. It was a teletype marked "Top Secret" from Washington, D.C. Grissom did not know anyone in Washington, but these orders were from the Pentagon. He was to report to a specific address in the city for a special meeting. He was not allowed to wear his uniform, or discuss these orders with anyone.

Grissom could not have predicted what would happen next. When he went to the meeting, he saw other men who looked a lot like him. They were test pilots wearing their best suits and ties. Later, Grissom was interviewed and tested and at the end of it all he found out what the meeting was about. He was invited to volunteer for Project Mercury. It did not take long for Grissom to give them his answer. "I knew instantly that this was for me," said Grissom.[4]

Grissom served in the space program with distinction. Of the first seven astronauts, he was one of three, along with Alan Shepard and John Glenn, to be chosen as a candidate to be the first American in space. That assignment was given to Shepard, but Grissom would be the second American in space in his capsule, *Liberty Bell 7.* Grissom's mission was a duplicate of Shepard's flight. He performed flawlessly as his capsule made the trip in and out of space in about fifteen minutes.[5]

However, when the capsule splashed down in the ocean, something went wrong. As Grissom was going through his postflight checklist, a small explosion blew the hatch off his capsule too soon. Water rushed into the tiny spacecraft and Grissom scrambled out, afraid he might drown. *Liberty Bell 7* sank to the ocean floor, and instead of getting a hero's welcome, Grissom got a lot of questions

*G*us Grissom was the second American in space, in the capsule, *Liberty Bell 7.*

about what happened in the moments after his spacecraft hit the ocean surface. It was determined, however, that it was not Grissom's fault that the hatch blew open.[6]

Ironically, the problem in Grissom's Mercury spaceflight became a contributing factor in the *Apollo 1* fire. Since it seemed that the hatch on *Liberty Bell 7* that blew open after splashdown was faulty, it was redesigned. Instead of opening outward, all of the new hatches would open inward and would be more securely affixed to the capsule.[7] Grissom resumed his career in the space program more successfully during Project Gemini. He served as commander on *Gemini 3*, a mission for which he named his spacecraft *Molly Brown* because he vowed that it would be unsinkable after hitting the ocean. (Molly Brown was a passenger aboard the *Titanic* who survived the ship's sinking. She later proclaimed herself "unsinkable.") Grissom was a key figure in gathering information that was very important in getting a man to the moon. He also would work at Mission Control in guiding his *Apollo 1* crewmate, Ed White, on one of the most exciting missions of the space program, *Gemini 4*.

Ed White's admission into the space program was a bit different from Grissom's. Like Grissom, White had experience as an Air Force fighter pilot during the 1950s,[8] but he was also a graduate of West Point, the prestigious military academy in upstate New York.[9] After getting a degree in aeronautical engineering at the University of Michigan, he applied for the new "rocket pilot" program at Edwards Air Force Base. Eventually, White was accepted

Gus Grissom is seen here with his wife, Betty, and their two sons, Scott (left) and Mark (right).

as one of nine new astronauts brought into NASA shortly after Wally Schirra's six orbits around the earth in 1962.[10]

Ed White made history for his participation on the *Gemini 4* mission. He was assigned to the flight with Jim McDivitt. Unlike his co-pilot, White would have to perform an EVA, or space walk. A space walk was originally planned for a later *Gemini* mission but the Russians had just sent one of their men on the first space walk. NASA wanted to meet the challenge as soon as possible. NASA decided that Ed White would be the first American to open up the door to his spacecraft and "take a walk" outside.[11]

The importance of the space walk was not only to keep up with the Russians. NASA needed to find out if an astronaut could survive in the vacuum of space, wearing only his space suit, if it was going to send a man to walk on the moon.

The *Gemini 4* spacecraft was launched on June 3, 1965. McDivitt piloted the spacecraft as White prepared for the EVA. Mission Control gave the "go" signal to exit the spacecraft. The world listened intently to the dialogue between White, McDivitt, and Gus Grissom, who

*G*rissom, White, and Chaffee enter the *Apollo 1* capsule during a launch rehearsal.

held the position of CAPCOM in Houston, communicating directly with the capsule.

"The view from up here is something spectacular," said White as he drifted away from the open hatch of his ship.[12] White's suit had twenty-one layers of material to protect him from the 250-degree heat from the sun's rays and the 150-degree-below-zero freeze of the spacecraft's shadow. He used a gun that shot pressurized oxygen to push himself out into open space and to maneuver his body around the ship. His "leash" extended twenty-five feet from the spacecraft, feeding him air to breathe and giving him the ability to talk to McDivitt and Grissom.[13]

Despite the very real danger of the space walk, White began to feel extremely happy, almost like a child on an amusement park ride. "This is fun!" he exclaimed as he looked at the world spinning three hundred miles a minute beneath him.[14] He could see with remarkable clarity the wakes of ships in the Pacific Ocean, railroads that cut across continents, and the outlines of major cities. It was even clearer than what he was used to seeing from airplanes that he had flown a lot closer to the earth's surface.[15]

Back on Earth, Grissom was getting concerned about White's intense happiness. He was afraid that he was getting a drunk feeling that scuba divers sometimes have at great depths and pilots sometimes have at high altitudes.[16] This was no place for an astronaut to lose his sense of judgement. "Gemini Four," said Grissom in a stern voice, "Get back in!"[17]

White did not want to return to his seat but something happened that reminded everyone of how dangerous the EVA was. A thermal glove that White had left on his seat became weightless and slowly drifted upward through the open hatch and away into the black infinite reaches of space.[18] McDivitt called out to White, "Come on let's get back in here before it gets dark."[19] He was making a small joke like a parent trying to coax a child away from the deep end of a swimming pool. Then McDivitt became very serious. "They want you back in *now*," he said.[20] "I don't want to come back in," White answered, "but I'm coming."[21] As he propelled himself back to the ship he added, "It's the saddest moment of my life."[22]

As sad as he was when the twenty-two-minute space walk ended, White and NASA had a lot to be happy about. White's space walk surpassed the Russian space walk in that he was able to control his movements in space, whereas the Russian had only tumbled around at the end of his tether.[23] White had proved that not only could a man survive outside of his spacecraft, but meaningful activities like building a space station, repairing a satellite, or walking on the moon were possible.[24]

Roger B. Chaffee was the youngest member of the *Apollo 1* crew, and would have been the youngest person to fly in space if the mission had been completed. He was born in Grand Rapids, Michigan, in 1935, and had always maintained a great interest in flying, partially because his father had been a pilot.[25]

After completing his bachelor's degree in aeronautical engineering at Purdue University, he enlisted in the navy and became one of the youngest pilots to fly a photo reconnaissance plane. In fact, he took some important photographs during the Cuban Missile Crisis, when the Soviet Union had introduced potentially dangerous nuclear weapons into Cuba that threatened the United States.[26]

Even though he had always been an excellent student, Chaffee worked even harder in the astronaut program. "Ever since the first seven Mercury astronauts were named, I've been keeping my studies up," said Chaffee. "At the end of each year, the Navy asks its officers what type of duty they would aspire to. Each year, I indicated that I wanted to train as a test pilot for astronaut status."[27]

In the fall of 1963, Chaffee achieved his goal. He was selected as one of fourteen pilots who made up NASA's third group of astronauts. This group also included Edwin "Buzz" Aldrin, who would be the second man on the moon, and Eugene A. Cernan, who was the last man to walk on the moon. During the course of his training, he worked with his future *Apollo 1* crewmates. He assisted Gus Grissom at the CAPCOM when he was speaking with Ed White.[28]

Chaffee never realized his dream of traveling in space and walking on the moon. *Apollo 1* was to be his first spaceflight, but the tragic fire claimed his life instead. His participation in the quest to reach the moon, however, did allow him to achieve another goal that he stated while he was a senior in high school. He wanted to have his name written in history books.[29]

Virgil "Gus" Grissom is buried at Arlington National Cemetery, just outside of Washington, D.C.

After perishing in the *Apollo 1* fire, all three astronauts were honored as national heroes. Grissom and Chaffee were buried in Arlington National Cemetery just outside of Washington, D.C. White had made it clear to his wife that if anything ever happened to him that he wanted to be buried at West Point.[30] Along with their families and NASA colleagues, the people of the nation mourned the deaths of the three heroes.

Once the astronauts had been laid to rest, NASA had to resume its mission to reach the moon. Before that could happen, NASA had to find out what caused the *Apollo 1* fire, and how to prevent it from happening again.

CHAPTER 4

What NASA Learned From the Fire

The deaths of the *Apollo 1* crew were devastating to many people. Some people who worked at NASA described the feeling as a "punch in the stomach" that left everyone feeling guilty that they had let the crew down.[1] The fire had also affected the nation's citizens as well. It reminded them that the United States was in the space race only because people were willing to risk their lives trying to get to the moon. The deaths of Grissom, White, and Chaffee had produced a period of national mourning, but soon after NASA would have to face the challenge of resuming the space program. Before NASA could do that, it had to learn exactly what happened in the *Apollo 1* fire.

"If there was anything that could be retrieved from this tragedy, it was the evidence—it was right there in front of us on Pad 34. We had a chance to discover the cause of the fire before another spacecraft was put at risk," said Gene Krantz, one of NASA's flight directors, many years after the fire.[2] In the days that followed the tragedy, astronaut Frank Borman was put in charge of the investigation of the *Apollo 1* fire. He and the rest of the investigation board set out to learn what the *Apollo 1* tragedy could teach them.

One of the contributing factors had nothing to do with anything on the launchpad, but rather with everyone

*T*he fire-damaged exterior of the *Apollo 1* command module. Because the fire occurred on the ground, the capsule could be examined and corrections could be made to later ships.

working at NASA that day. "Go fever" had set in. The Mercury and Gemini programs had gone so well, that success had begun to seem routine. People were eager to get the first mission of Project Apollo under way. Everyone worked quickly and had become accustomed to putting up with faulty hardware and poor workmanship. That day, most of the people had been working since the early morning and had encountered problems all day long. Nearly everyone was tired, maybe a little bored as they neared the end of the routine "plugs out" test, and not mindful of the extremely dangerous working conditions.[3]

One of the most dangerous contributing factors to the *Apollo 1* fire was the atmosphere of pressurized oxygen in the command module. Once Grissom, White, and Chaffee had boarded the cockpit, they were sealed inside. The capsule was then filled with pure oxygen. In order to test the spacecraft for leaks, the oxygen was pressurized up to approximately 15 p.s.i.'s (fifteen pounds of force per square inch). In an atmosphere such as this, aluminum burns easily. That explained why most of the spacecraft's materials ignited so quickly.[4]

The oxygen's pressure put the astronauts in additional danger. "When the capsule was pressurized there was no way the hatch would open," recalled astronaut Pete Conrad.[5] Ed White had struggled to open the hatch, in the midst of flames and poisonous gas, against an impossibly strong force.

The poor construction of the spacecraft was another contributing factor. The *Apollo 1* spacecraft had many

design flaws and sloppy
workmanship. One of the
worst parts of the command
module was the wiring, and
it is believed that a short cir-
cuit in the poorly insulated
wiring was the cause of
the fire.[6]

*T*he burned interior of *Apollo 1* command module.

After the investigation,
NASA made some changes
in all of its missions to
follow. More than one
hundred design changes were made in the Apollo spacecraft. These changes included a hatch that would open
outward in five seconds, better shielding on the wiring,
better fireproofing, and no more ground tests in an atmosphere of pure oxygen.[7]

The Apollo spacecraft continued to use an atmosphere
of pure oxygen during spaceflight, but the materials
inside the spacecraft were carefully tested for flammabili-
ty. Today, the space shuttles and the International Space
Station are designed to use a mixture of nitrogen and oxygen to reduce the danger of fire.

While nothing could replace the loss of life, NASA did
inherit some blessings from the tragedy. Chris Kraft
would have been the flight director of *Apollo 1*. Many
years later he said in television interview, "I don't think
we would have gotten to the moon in the sixties if we had
not had the fire. That's a terrible thing to say, but I think

it's true." [8] Rocco Pettrone thought about how the incident could have been worse. "If that had happened while [astronauts] were on the way to the moon, we would have lost the crew, never heard from them again, [and] there would have been a mystery hanging over the whole program which would have caused an untold delay, and maybe even a cancellation [of putting a man on the moon]." [9] Because the *Apollo 1* fire occurred on the ground, all of the wreckage could be analyzed and the incident reviewed in order to find out what had gone wrong.

CHAPTER 5

Project Apollo Takes Flight Again

The tragedy of *Apollo 1* caught NASA by surprise, taught some painful lessons, but finally left the space program with the challenge of resuming its mission to the moon. Project Apollo continued with a series of unmanned missions that studied the surface of the moon. On October 11, 1968, Commander Walter Schirra and his crew, Walt Cunningham and Don Eisele, tested the new spacecraft on *Apollo 7.*[1] Their mission successfully completed the tasks that were to be done by *Apollo 1.* However, many of Project Apollo's most remarkable moments were still yet to come.

Perhaps the most amazing Apollo mission, other than putting men on the moon, was the mission of *Apollo 8.* The goal of *Apollo 8* was to leave the earth's orbit, travel to the moon, orbit the moon a number of times, and return home. None of these had ever been accomplished by a manned mission before.[2] However, the flight plan did not reveal to anyone, not even the crew of Frank Borman,

*T*he lunar module of *Apollo 11* slowly descends toward the surface of the moon. *Apollo 11* landed on the moon seven months after *Apollo 8* completed an orbit of the moon.

Jim Lovell, and Bill Anders, what amazing things would happen on this journey.

Once the *Apollo 8* spacecraft had reached the moon, the crew prepared to continue to the far side. In what Jim Lovell would later call "the longest four minutes I've ever spent," the spacecraft was separated from the earth, breaking radio contact with home.[3]

In a precise moment that had been calculated by NASA scientists, the windows of the spacecraft were suddenly brightly lit as they came back around to the lighted side of the moon. The *Apollo 8* astronauts saw the surface of the moon up close for the first time. "We were like school kids looking into a candy store window. Our noses were pressed against the glass," said Lovell.[4] The moon was just a short distance beneath them showing a lifeless and desolate, yet magnificent, terrain. As they looked at the moon wondrously, the astronauts gave temporary names to the biggest craters in honor of all those who helped them reach their destination. They named three of the craters Grissom, White, and Chaffee.[5]

Hanging in the vast, empty darkness was home, the planet Earth. It looked small and fragile. The astronauts had been instructed to take photographs of the moon's surface in order to find a landing site for the missions that

*A*stronauts aboard the *Apollo 11* answered President Kennedy's challenge when they walked on the moon on July 20, 1969.

would follow. However, they found themselves also taking many photos of the earth. Bill Anders later remembered, "Here we came all this way to the moon and the most significant thing we're seeing is our own home planet, the earth."[6] The *Apollo 8* photos would give the people of the entire planet their first real look at Earth. One of the pictures was even made into a United States postage stamp. It showed the earth hanging over the moon's horizon and it was called "earthrise."

At the time the *Apollo 8* astronauts were taking their photographs, the earth itself and the United States in particular seemed to be coming apart. It was December 1968, and the war in Vietnam was at its height. The war was just one of many issues that had divided people in the United States during a very violent year. As American soldiers died in Vietnam, protests erupted in the United States including outside the Democratic National Convention during the summer of 1968. Earlier in the year the civil rights movement lost one of its most inspirational leaders when Dr. Martin Luther King, Jr. fell to an assassin's bullets. Senator Robert F. Kennedy also was assassinated while he was running for president. His death not only reminded the nation of President Kennedy's assassination nearly five years earlier, but also added to the growing feeling of hopelessness in the nation. That was the world that the crew of *Apollo 8* had left behind, but it did not look that way through their windows. It looked peaceful, fragile, and beautiful.

Before the launch, NASA had told the *Apollo 8* astronauts they had to address the world from their spaceship during their mission. It happened to be Christmas Eve, and they were told to "do something appropriate." With the largest audience ever to hear a human voice from space, Bill Anders spoke to his listeners from the spacecraft.

"For all the people on earth, the crew of *Apollo 8* has a message we would like to send you," said Anders. He began to read, "In the beginning, God created the heaven and the earth and God said let there be light: and

An *Apollo 1* patch like the one left on the moon by Buzz Aldrin.

there was light and it was good." He was reading from Genesis, the first book of the Bible. The message surprised many people back on the ground. Jim Lovell read the next four verses. Commander Frank Borman finished the passage and then closed by saying, "and from the crew of *Apollo 8*, we close with good night, good luck, a Merry Christmas, and God bless all of you—all of you on the good earth."[7]

The words were directed to everyone of the planet. It was a beautiful message of hope during a very confusing time.

The culminating mission of Project Apollo was *Apollo 11*, when Neil Armstrong and Edwin "Buzz" Aldrin walked on the moon on July 20, 1969. President Kennedy's challenge had been met, with a few months to spare. All of the knowledge NASA had gained during Projects Mercury and

Gemini as well as all of the previous Apollo missions, including the *Apollo 1* fire, had been applied to put men on the moon and bring them home. Grissom, White, and Chaffee had not given their lives in vain. In tribute to his friends who died in the fire, Buzz Aldrin left an *Apollo 1* patch on the moon.

Project Apollo did not end with the first moon walk. Other missions followed, including the infamous voyage of *Apollo 13.* Then a disaster was narrowly averted when the spaceship blew an oxygen tank, and the crew had to come home in the tiny lunar module. *Apollo 17* was the last manned mission to the moon. Its commander,

*T*he *Apollo 13* is retrieved shortly after making splashdown.

Eugene Cernan, holds the distinction of being the last man to walk on the moon.

Since it began in 1958, NASA has launched numerous missions into space with comparatively little tragedy. The grief the country knew following the *Apollo 1* fire was briefly revisited by the explosion of the space shuttle *Challenger* in 1986. Seven astronauts perished, including a civilian schoolteacher, Christa McAuliffe, minutes after lift off. Just as with the *Apollo 1* fire, the accident was investigated, and lessons were learned and applied to the further exploration of space.

Now NASA is working with other nations to complete the International Space Station (ISS), which is being constructed in the earth's orbit. The goal of the ISS is to unite the countries of the earth in the cooperative

*T*he space shuttle *Challenger* takes off on its fatal journey on January 28, 1986.

*T*he seven-person crew of *Challenger*: Mike Smith, Dick Scobee, Ron McNair (front row, left to right); Ellison Onizuka, Christa McAuliffe, Gregory Jarvis, and Judith Resnik (back row, left to right).

exploration of space in order to benefit the entire human race. Scientists hope to gain a better understanding of the effects of gravity, as well as pave the way for private business to engage in commercial opportunities in space.[8]

The International Space Station carries on the legacy of those who were willing to make the greatest sacrifices to explore space. Gus Grissom surely understood the great risks, and the great triumphs, that were a part of space travel. He said once during an interview, "If we die, we want people to accept it. We're in a risky business, and we hope that if anything happens to us it will not delay the program. The conquest of space is worth the risk of life."[9]

Other Spaceflight Disasters

DATE	PLACE	EVENT
October 1960	Baykonur, Soviet Union	After an R-16 rocket's engines fail to ignite for launch, a team of Soviet engineers is ordered to ignore procedure and immediately inspect the rocket on the launchpad, before pumping away the fuel. An explosion blows apart the fuel tanks and a fire quickly breaks out. Nearly a hundred people are killed, including several top engineers of the Soviet space program.
December 1965	United States	A stuck fuel valve halts the launching of the *Gemini 6*. Gambling that there is no danger of an explosion, astronauts Wally Schirra and Tom Stafford do not eject, allowing the fuel to be safely pumped away. The launch is successfully completed three days later, allowing the *Gemini 6* to make its appointed rendezvous with the *Gemini 7*.
1967	Soviet Union	*Soyuz 1* is launched with a single pilot. It is scheduled to link up with a second manned spaceship, but *Soyuz 1* develops problems and is ordered to return to the earth. *Soyuz 1* crashes due to parachute failure, killing the pilot, Vladimir Komarov.
November 1969	United States	*Apollo 12* experiences technical difficulties during launch, possibly due to being struck by lightning shortly after leaving the ground. The craft goes dark for several seconds before backup power comes on. The *Apollo 12* team successfully completes its mission and returns safely to Earth.
April 1970	United States	About fifty-six hours after the beginning of its flight, a short circuit causes an explosion aboard *Apollo 13*, destroying the command module's life support and electrical systems. The three-man crew retreats into the lunar module, which has enough oxygen to keep them alive while they circle the moon and return to Earth. The crew makes a safe splashdown in the command module with the lunar module still attached.
January 1986	United States	The space shuttle *Challenger* explodes almost immediately after liftoff as a result of propellant leaking through a faulty seal. The entire seven-person crew of the *Challenger*, including schoolteacher Christa McAuliffe, is killed in the disaster.

Chapter 1. The Strange Day of January 27, 1967

1. *Space Flight*, Signature Communications Co. for WETA/Washington & WYES/New Orleans, television documentary, 1986.

2. Ibid.

3. Courtney G. Brooks, James M. Grimwood, and Loyd S. Swenson, Jr., *Chariots for Apollo*, (Washington, D.C.: National Aeronautics and Space Administration Scientific and Technical Office,1979), p. 213.

4. Alan Shepard and Deke Slayton, *Moon Shot* (Atlanta, Ga.: Turner Publishing, Inc., 1994), p. 196.

5. Buzz Aldrin and Malcolm McConnell, *Men From Earth* (New York: Bantam Books, 1989), p. 162.

6. Brooks, Grimwood, and Swenson, p. 214.

7. Shepard and Slayton, p. 196.

8. Ibid., p. 198.

9. Ibid., p. 196.

10. Gene Krantz, *Failure Is Not An Option* (New York: Simon & Schuster, 2000), p. 196.

11. Ibid., p. 197.

12. Shepard and Slayton, p. 200.

13. Aldrin and McConnell, p. 167.

14. Shepard and Slayton, p. 204.

15. Ibid., p. 205.

16. Ibid., p. 204.

17. *Moon Shot*, television documentary, TBS Productions, 1994.

18. Brooks, Grimwood, and Swenson, p. 217.

Chapter 2. The Space Race

1. Buzz Aldrin and Malcolm McConnell, *Men From Earth* (New York: Bantam Books, 1989), p. 161.

2. *Space Flight*, Signature Communications Co. for WETA/Washington & WYES/New Orleans, television documentary, 1986.

3. Ibid.

4. Alan Shepard and Deke Slayton, *Moon Shot* (Atlanta, Ga.: Turner Publishing, Inc., 1994), p. 49.

5. Ibid, p. 45.

6. *Space Flight*, television documentary, 1986.

7. Shepard and Slayton, p. 96.

8. Gene Krantz, *Failure Is Not An Option* (New York: Simon & Schuster, 2000), p. 55.

9. John F. Kennedy, *Special Message to the Congress on Urgent National Needs*, May 25, 1961, <http://www.jkflibrary.org/j052561.htm> (August 21, 2001).

10. *Moon Shot*, television documentary, TBS Productions, 1994.

Chapter 3. The Three Men of Apollo 1

1. Loudon S. Wainwright, "Grissom: A Quiet Little Fellow Who Scoffs At The Chance Of Becoming A Hero," *Life*, March 3, 1961, p. 28.

2. Virgil "Gus" Grissom, *Gemini!* (New York: The Macmillan Company, 1968), p. 18.

3. Grissom, pp. 18–19.

4. Grissom, p. 22.

5. Alan Shepard and Deke Slayton, *Moon Shot* (Atlanta, Ga.: Turner Publishing, Inc., 1994), p. 141.

6. Ibid., p. 142.

7. Ibid., p. 196.

8. Buzz Aldrin and Malcolm McConnell, *Men From Earth* (New York: Bantam Books, 1989), p. 31.

9. Ibid., p. 8.

10. Shepard and Slayton, p. 157.

11. Ibid., p. 181.

12. Ibid., p. 182.

13. Ibid., p. 181.

14. Ibid., pp. 181–182.

15. Ibid., p. 182.

16. Ibid.

17. Ibid.

18. Ibid., pp. 181–182.

19. Gene Krantz, *Failure Is Not An Option* (New York: Simon & Schuster, 2000), p. 140.

20. Shepard and Slayton, p. 182.

21. Ibid.

22. Aldrin and McConnell, p. 128.

23. Grissom, p. 125.

24. Ibid., p. 126.

25. Mary C. Zornio, "Detailed Biographies of the Apollo 1 Crew-Roger Chaffee," *NASA History*, 1997, <http://www.hq.nasa.gov/office/pao/History/Apollo204/zorn/chafee.htm> (August 21, 2001).

26. Ibid.

27. Ibid.

28. Ibid.

29. Ibid.

30. Shepard and Slayton, p. 212.

Chapter 4. What NASA Learned From the Fire

1. *To The Moon*, television documentary, A Nova Production, Lone Wolf Pictures for WBGH/Boston, 1999.

2. Gene Krantz, *Failure Is Not An Option* (New York: Simon & Schuster, 2000), p. 200.

3. *Moon Shot*, television documentary, TBS Productions, 1994.

4. *To The Moon*, television documentary, 1999.

5. Ibid.

6. Ibid.

7. Ibid.

8. Ibid.

9. Ibid.

Chapter 5. Project Apollo Takes Flight Again

1. Alan Shepard and Deke Slayton, *Moon Shot* (Atlanta, Ga.: Turner Publishing, Inc., 1994), p. 221.

2. *Moon Shot*, television documentary, TBS Productions, 1994.

3. Shepard and Slayton, p. 231.

4. *To The Moon*, A Nova Production, Lone Wolf Pictures for WBGH/Boston, 1999.

5. Gene Krantz, *Failure Is Not An Option* (New York: Simon & Schuster, 2000), p. 245.

6. *To The Moon*, television documentary, 1999.

7. Shepard and Slayton, p. 233.

8. "NASA Expedition One Press Kit," *NASA Human Spaceflight*, October 25, 2000, <http://www.spaceflight.nasa.gov/station/> (August 21, 2001).

9. Virgil I. Grissom, "Apollo 1 Mission: In Memoriam," *NASM—Apollo to the Moon*, March 1965, <http://www.nasm.edu/galleries/attm/rm.br.ea.4.html> (August 21, 2001).

CAPCOM—This is short for "capsule communicator" and is the only station in Mission Control that communicates directly with the astronauts in the capsule.

Extra Vehicular Activities (EVAs)—"Space walks" to test an astronaut's ability to survive in space with only a space suit for protection.

Kennedy Space Center—The field center based in Florida that is in charge of preparing and launching a crew and spacecraft for its launch.

Mission Control—The field center based in Houston, Texas, that is in charge of guiding a crew and spacecraft through its mission after launch.

National Aeronautics and Space Administration (NASA)—The official agency of the federal government in charge of space exploration.

"plugs out" test—A dress rehearsal for the launch of a rocket.

Project Apollo—The third and final phase of NASA's effort to reach the moon, which included a series of three-man missions.

Project Gemini—The second phase of NASA's effort to reach the moon, which included a series of two-man missions.

Project Mercury—The first phase of NASA's effort to reach the moon, which included a series on one-man missions.

rendezvous—A maneuver in which two spacecraft link together in space.

Bredeson, Carmen. *The Challenger Disaster: Tragic Space Flight.* Berkeley Heights, N.J.: Enslow Publishers, Inc., 1999.

Bredeson, Carmen. *Gus Grissom: A Space Biography.* Berkeley Heights, N.J.: Enslow Publishers, Inc., 1998.

Bredeson, Carmen. *The Moon: A First Book.* Danbury, Conn.: Franklin Watts, Inc., 1998.

Chaikin, Andrew. *A Man on the Moon: The Voyages of the Apollo Astronauts.* Volumes 1–3. Alexandria, Va.:Time-Life, Inc., 1999.

Cole, Michael D. *Space Launch Disaster: When Liftoff Goes Wrong.* Berkeley Heights, N.J.: Enslow Publishers, Inc., 2000.

De Angelis, Gina. *The Apollo 1 and Challenger Disasters.* Broomall, Pa.: Chelsea House, 2001.

Kelly, Nigel. *The Moon Landing: The Race into Space.* Chicago, Ill.: Heinemann Library, 2001.

Kennedy, Gregory P. *Apollo to the Moon.* Broomall, Pa.: Chelsea House, 1992.

Spangenburg, Ray and Kit Moser and Diane Moser. *Project Apollo.* Danbury, Conn.: Franklin Watts, Inc., 2001.

Internet Addresses

The Official NASA Web Site
http://www.nasa.gov

The Official NASA History Web Site
http://history.nasa.gov

The Kennedy Space Center
http://www.ksc.nasa.gov

The Smithsonian Institution National Air & Space Museum
http://www.nasm.si.edu